ABACUS 101

ABACUS 101

Welcome to the world of abacus.

BEGINNER ABACUS MATH by Tong Dazai

Abacus 101 Series Copyright © 2017 by David Dong, All Rights Reserved

Illustrations Copyright © 2017 by Tong Dazai, All Rights Reserved.

A Fun Educational Guidebook.

An Imprint of West Covina, California.

Printed in the United States of America by Amazon Inc., Seattle WA.

Abacus 101 Series books may be purchased for business or promotional use.

For more information on bulk purchases, please contact daviddong012890@gmail.com.

ISBN: 978-1-520-22564-7 (Hardcover)

ALL RIGHTS RESERVED

Beginner Abacus Math

A Fun Way To Learn Basic Math (Entry Level)

An Abacus 101 Series By Tong Dazai

Table of Contents

What is an Abacus and why is it good for me or my child? 9
    How long does it take to learn Abacus? 10
Chapter 1: The Story of Abacus 11
Chapter 2: Place Values & Number Values 13
Chapter 3: "Clear" to Zero 15
Chapter 4: Adding 1 to 9 17
    Double Digits 20
Chapter 5: Visual Practice 22
    Chapter Review 33
Chapter 6: Rules of Addition & Subtraction 34
Chapter 7: Thumb up & Index down 36
Chapter 8: Top bead & Index down 37
Chapter 9: Top bead & Index up 38
Chapter 10: Thumb + Index additions 39
Chapter 11: Thumb + Index additions 40
Chapter 12: Thumb + Index additions 41
Chapter 13: Thumb + Index subtractions 42
Chapter 14: Thumb + Index subtractions 43
Chapter 15: Thumb + Index subtractions 44

What is an Abacus and why is it good for me or my child?

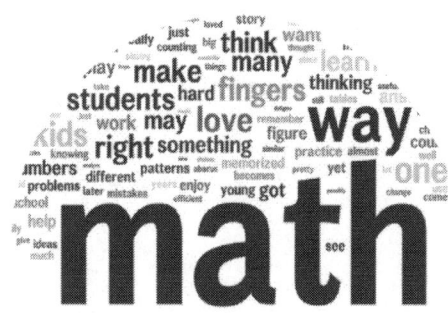

"But in my opinion, all things in nature occur mathematically."

Have you ever seen children on YouTube calculate faster than a calculator? Do you want to be able to know what it feels like to be super confident in math? Do you just want to expand your general arithmetic knowledge and reasoning? Well, learning the basics of how to use an abacus can help you achieve all of these, and in a shorter amount of time than you might have thought!

At its very simplest, abacus is a language just like you'd read aloud from a book. The beads you'll see on the abacus are like the musical notes on a score sheet. And they each represent a different number and place value, as well as groups and patterns such as the composition of fives and tens. The method of an abacus has been used for hundreds of years and is the symbol of wisdom in numbers. Learning to read numbers on an abacus really does open up a whole new world to explore!

The physical movement of the abacus beads will imprint a mental abacus inside your mind, which will allow you to mentally exercise your mind. By stimulating, abacus can help your brain in concentration, observation, memory, comprehension, logical thinking, critical thinking, problem solving, fast judgment, endurance, determination, confidence and the general interest in numbers. It really feels like a cup of warm coffee on a cool day.

Our technique and method of training has helped our students in overcoming their difficulties and fear in numbers. The interactive program has developed many "math-genius" and you are about to know the secrets behind it. The importance of math in this world is inevitable, and abacus can help you climb faster on the ladder to a greater knowledge and wisdom in the arithmetic world. Abacus with us will be a new fun experience that will help you or your child tremendously in life.

How long does it take to learn Abacus?

Similar to how traditional schools are divided into elementary, junior high, high school, university and grad school, abacus is divided into levels.

The abacus level is a method of demonstrating your skill, and the higher the level the harder it is. Below is the overall breakdown of each level from easy to hard:

- *Entry Level - Learn to identify the numbers on an abacus, basic addition and subtraction and memorize abacus rules (THIS BOOK)*
- *Level 10 - Basic mental math*
- *Level 9 - Double digit addition & subtraction*
- *Level 8 - Multiplication & division*
- *Level 7 - Single and double digit numbers*
- *Level 6 - Multiple digit multiplication & division*
- *Level 5 - Focus, speed and accuracy*
- *Level 4 - Super mental math*
- *Level 3 - Supernatural concentration*
- *Level 2 - Dollar and cents*
- *Level 1 - 5 digit abacus and mental calculation*
- *Level Degree - 7 digit abacus mental calculation*

Everyone who learns abacus can learn at their own pace. It is meant to be a fun experience and exercise for you or your child. In this book, you will learn how to read and identify the numbers on the abacus instrument. It will guide you to a fundamental understanding of how an abacus works, including how to add & subtract and the concepts behind it all.

"Math feels easier and it makes sense." -Ashley Espada, age 9.
"I see numbers and I can move them in my head. It's fun!" -Carter Chu, Age 12.
"Math is not scary anymore." -Jasmine Wang, Age 10.

Regardless of age, abacus will train your left and right brain to work together as a team, which will benefit you for a lifetime. Let's begin!

Chapter 1: The Story of Abacus

It is difficult to imagine counting without numbers, but there was a time when written numbers did not exist. The earliest counting device was the human hand and its fingers, capable of counting up to 10 things; toes were also used to count in tropical cultures.

Then, as even larger quantities (greater than ten fingers and toes could represent) were counted, various natural items like pebbles, sea shells and twigs were used to help keep count.

Merchants who traded goods needed a way to keep count (inventory) of the goods they bought and sold. Various portable counting devices were invented to keep tallies.

The abacus is one of many counting devices invented to help count large numbers. When the Hindu-Arabic number system came into use, abaci were adapted to use place-value counting.

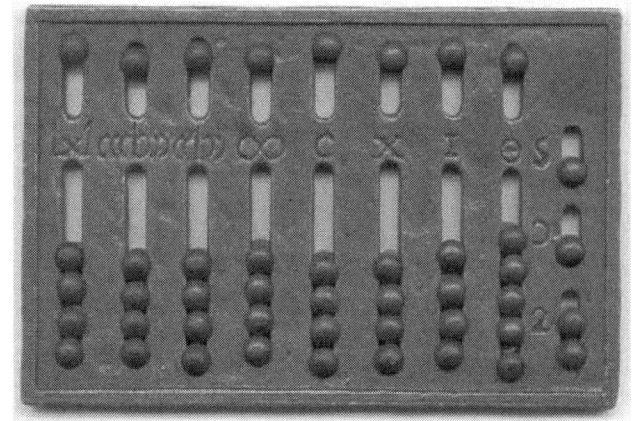

It is important to distinguish the early abacuses (or abaci) known as counting boards from the modern abaci. The counting board is a piece of wood, stone or metal with carved grooves or painted lines between which beads, pebbles or metal discs were moved.

The abacus is a device, usually of wood (romans made them out of metal and they are made of plastic in modern times), having a frame that holds rods with freely-sliding beads mounted on them

The abacus, called *Suan-Pan* in Chinese, as it appears today, was first chronicled in 1200 C.E. in China. Rules were created so people are able to do complicate arithmetic calculations on the abacus. In 1900s, abacus was a tool that was used daily by mathematicians, accountants and merchants. Today, abacus is widely popular in Asia as a development program for children and adults.

Personally, abacus feels like a magical wand that is able to move numbers wherever-as-I-like. The ones, fives and tens all work together to help me reach my answers quicker than others. Math feels like my second language, natural and makes sense. My brain is able to stay healthy and I am able to absorb new concepts and ideas.

Chapter 2: Place Values & Number Values

In our decimal number system, the value of a digit depends on its place, or position, in the number. Each place has a value of 10 times the place to its right. A number in standard form is separated into groups of three digits using commas. Each of these groups is called a period. On the abacus, the system works the same way and it's visually easy to see.

This is a standard 17 digit abacus. Each bead on the abacus represents a number. The period on the abacus represents the **ones** place. The position to the left of the ones place will be tens, hundreds, thousands and so on. You can start on any period on the abacus, however, we recommend starting in the middle period as it is easier to see.

To get started, find the middle period on the abacus. Each bead (or stone) on the bottom of the shaft equals to 1, and there are four 1s on the bottom so you can use them to create 1, 2, 3 or 4. The bead on the top of the shaft represents 5. By combining 5 and 1s, ones place value can go up to 9.

The same concept applies to the tens place. In tens place, each bead on the bottom of the shaft equals 10, and there are four 1s so you can use them to create 10, 20, 30 or 40. The 5 on the top of the shaft represents 50. The combinations are limitless as there is an infinite amount of numbers.

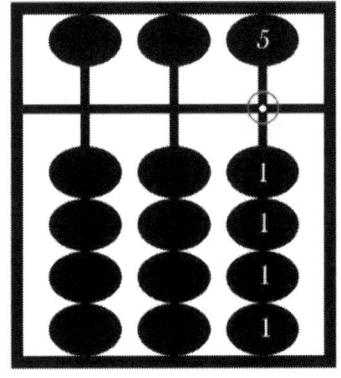

Each shaft represents a placement

Stone on top equals 5

The dot represents ONES place

Each stones on the bottom equals 1

Combination of stones can create numbers. Right now is 0.

*This is the "CLEAR" position.

On the image, the dot that is circled in red represents the **ones** place. Here you can see each of the bottom beads are labeled as 1.

Right now, the abacus is at zero. To add or subtract, you will have to move the beads around the shaft to make combination of numbers. The bead at the top of the red circle is labeled as 5. The maximum value of the **ones** place value can go up to 9.

Chapter 3: "Clear" to Zero

The first thing we have to do on the abacus before we start is to clear the abacus. By clearing, it will reset the abacus back to zero. In this picture of the abacus, it is ready to be cleared. The abacus must be cleared before we can start to work on it, otherwise the beads are everywhere and it will be impossible to tell what is the answer.

1. Lay the abacus flat on the desk.
2. With the off hand, hold firmly on the end of the abacus.
3. Gently tilt the abacus in an upward position.
4. Once the beads falls down, slowly lay the abacus back down.
5. Place your index finger beneath the top bead.
6. With your index finger out and other fingers tucked in, start by placing your index finger beneath the 5 and slide it across the abacus.

This is how a cleared abacus should look like. The top and the bottom beads are neatly separated by the middle shaft. This is the same as the clear function on a calculator, and must be performed before you start your work.

Tip: It is essential to be comfortable with this action as well as the correct finger posture.

Practice the clearing motion several times by shaking the beads up on the abacus and clear them afterward. It may feel uncomfortable to your finger at first, but with practice, clearing should be very smooth and takes less than a second to do.

There are several common errors you want to avoid when performing the clear action on the abacus. First error is letting go of your off-hand. This will result in the instability of the abacus, and your abacus will shake and the beads will be out of place. Make sure your off-hand is on the abacus the whole time you are working on it to be sure it is stable.

The second common error is when students are clearing the abacus and they feel like the beads are stuck and they use excessive force the make the beads go up.

Clearing should be very smooth and gentle to the abacus. Make sure your index finger is properly placed underneath the top bead and gently slide across the top in an outward motion. It may help when you use the nail part so it has less traction and will make clearing easier.

The third common error is when the student becomes too violent with the beads because it does not move the way they want. With practice, the beads will move the way you want without any excessive force, it just takes practice.

Be sure to be able to perform the clear action really well before moving onto the next chapter, which will be the number recognition.

Tip: Use the nail part of your index finger to slide beneath the top beads. It will make the motion smoother.

Tip: Practice until you are able to perform the clear action in a second or less. It is the most used action when using an abacus. Each time you start a question, you will need to perform "clear".

Chapter 4: Adding 1 to 9

In this chapter, you will learn how to properly add 1 to 9 on the abacus. In abacus, you will only need two fingers, which are your thumb and index finger. First, hold the abacus firmly with your off-hand at the end of the abacus. This will stabilize your abacus so it will not shake when you are moving the beads around. Your writing hand will be the hand to perform all the actions. To correctly position your hand, hold it out as a fist, then, have your thumb and index finger out and the rest of the fingers tucked in.

Find the period on your abacus. With your thumb, push 1 bead up so it looks like the image on the left.

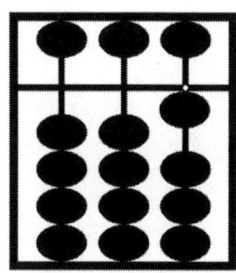

Lesson 1.
Add 1 with Thumb Up
Subtract 1 Index Down

On the abacus, it reads as 1. To subtract 1 on the abacus, use your index finger and push the bead back down. The abacus should now be at zero. Make sure you are not moving any other beads during your work, otherwise the answer will be incorrect.

Tip: Use the correct finger movement will help with the accuracy and the speed.

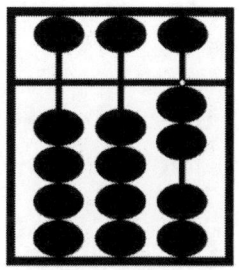

Lesson 2.
Add 2 with Thumb Up
Subtract 2 Index Down

With your thumb, add 2 by pushing 2 beads up at once. Remember not to push the beads up separately, otherwise you are just taking an extra step. Subtract 2 by both beads down together with your index finger.

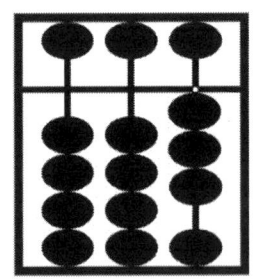

Lesson 3.
Add 3 with Thumb Up
Subtract 3 Index Down

Add 3 by pushing 3 beads up with your thumb. Again, this should be done in one motion. Subtract by pushing the beads down with your index finger.

Tip: A common error is when the beads are added separately, meaning when you add 3, you're adding 1+1+1, which will be incorrect.

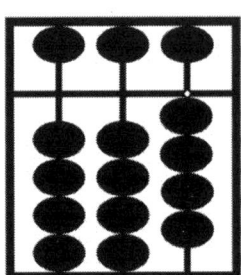

Lesson 4.
Add 4 with Thumb Up
Subtract 4 Index Down

Add 4 by pushing 4 beads up with your thumb. Subtract 4 by pushing the beads down with your index finger. Once more, remember to move all 4 beads together at the same time.

Tip: Clear the abacus every time when the beads look too shook up. This way you can practice clearing as well.

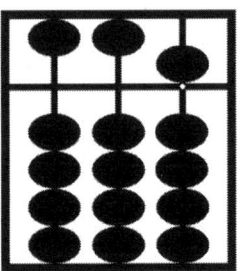

Lesson 5.
Add 5 with Index Down
Subtract 5 Index Up

Remember from the previous chapter, each bead on the top of the shaft represents 5. Add 5 by using your index finger and push the top bead down. To subtract 5, take away the top bead with your index finger.

It will feel differently than adding and subtracting the beads on the bottom. Practice 3 to 5 times until your finger movements are comfortable.

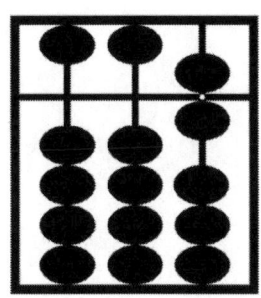

Lesson 6.
Add 6 with Thumb Up & Index Down
Subtract 6 Thumb Down & Index Up

Add 6 by using the combination of 5 and 1 together, making the number 6. Push the top bead down with your index finger and push the bottom bead up with your thumb. To do this, think of it as you are picking up something by pinching it. To subtract, do the opposite by taking away both beads with the same finger you've added with, and open them up.

Tip: Make it easier by placing your finger at the place you want to move the beads first, then perform the action. Think first before you move.

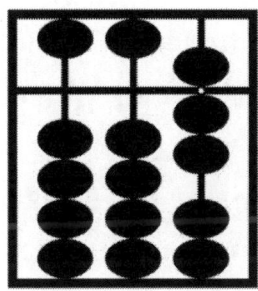

Lesson 7.
Add 7 with Thumb Up & Index Down
Subtract 7 Thumb Down & Index Up

Adding 7 is the same as adding 6, use both of your index and thumb to move the beads in a pinching motion at the same time. Remember to move the two bottom beads together. Subtract 7 by opening the top and bottom up with your thumb and index finger.

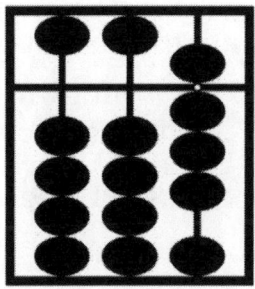

Lesson 8.
Add 8 with Thumb Up & Index Down
Subtract 8 Thumb Down & Index Up

Add 8 by using both of your thumb and index finger and add them together at the same time. Subtract 8 by opening it up with both fingers. Anytime the abacus becomes too messy, reset by performing the clear action.

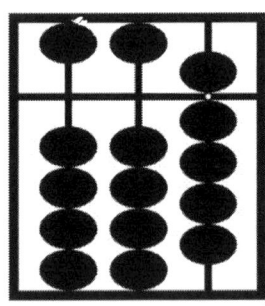

Lesson 9.
Add 9 with Thumb Up & Index Down
Subtract 9 Thumb Down & Index Up

Add 9 is the same method as 6, 7 & 8. Add 1 bead on top and 4 beads on bottom together at the same time. Make sure to pinch them together in one single motion.

Subtract by opening up your finger from the inside of the beads. This is the biggest digit you can have on the **ones place value.**

Double Digits

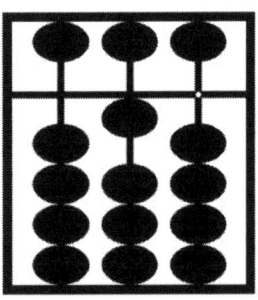

Lesson 10.
Add 10 with Thumb Up
Subtract 10 with Index Down

To add 10 on the abacus is straight forward. Find the place value of the tens place, and push the bead up. The picture to your left shows the tens place is left to the ones place.

By adding 1 bead on the tens place, you are adding a ten. Use the same finger movement as adding on the ones place value. Use your thumb and push up the bead. Subtract by pushing the bead down with your index finger.

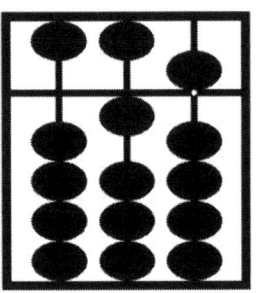

Lesson 11.
Add 15 with Thumb Up & Index Down
Subtract 15 with Index Down & Index Up

You can try your own combination of numbers. There is no limit on how big or small the number can be. For instance, adding 15 means having a 1 in the tens place and having a 5 in the ones place.

Adding 2 beads in the tens place will equal to 20. 3 beads will equal to 30 and so on. Add 50 by adding the top bead on the tens place. Add 60 by adding a 6 in the tens place.

To add 2 digits, add in the order from left to right. For instance, when you are adding 15, add 10 first then add 5. Here are a few numbers you can practice yourself by adding and subtracting: 18, 23, 41, 64, 59, 73, 88, 99, 105, 625.

Tip: Remember to add from left to right.

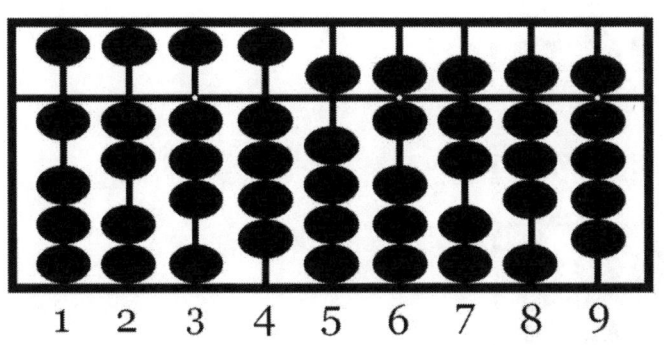

You can practice adding 1 to 9 by starting off at a period and add 1 to 9 from left to right. Subtract 1 to 9 in the order it was added. This way you can practice all 9 digits and have an overview of how each number looks like on the abacus.

Chapter 5: Visual Practice

In this exercise, write down the numbers you see on the abacus in this booklet or on a separate sheet of paper if you are reading the ebook.

Are you ready?

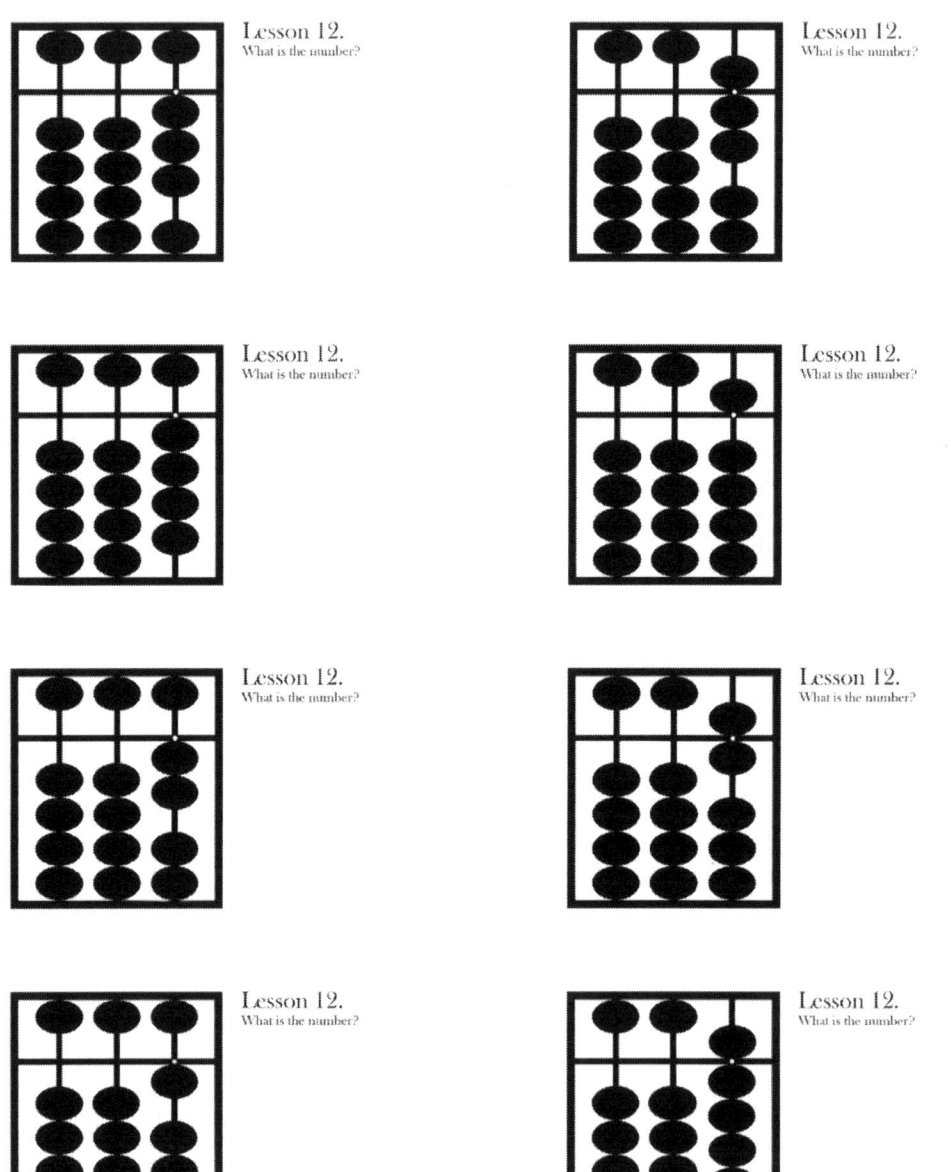

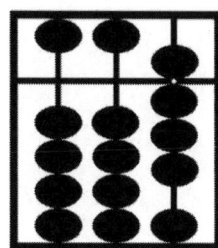

Lesson 12.
What is the number?

The answers: 3, 7, 4, 5, 2, 6, 1, 9, 8. Did you get it correct? Now, let's make it a little bit harder. Write out the numbers you see.

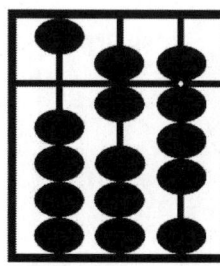

Lesson 13.
What is the 2 digit number?

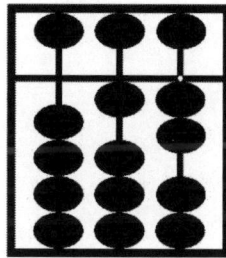

Lesson 13.
What is the 2 digit number?

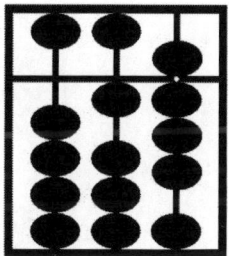

Lesson 13.
What is the 2 digit number?

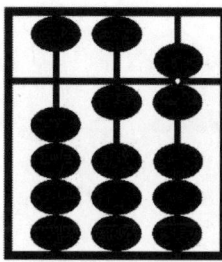

Lesson 13.
What is the 2 digit number?

Lesson 13.
What is the 2 digit number?

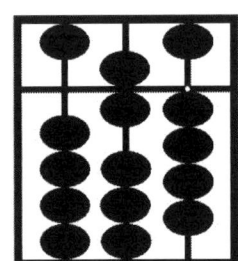 Lesson 13.
What is the 2 digit number?

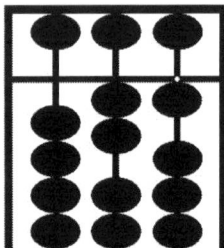 Lesson 13.
What is the 2 digit number?

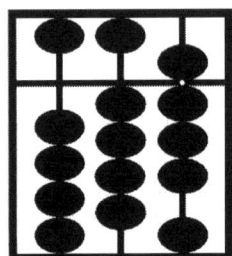 Lesson 13.
What is the 2 digit number?

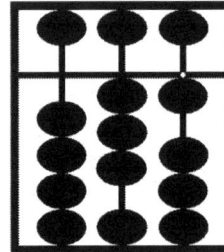 Lesson 13.
What is the 2 digit number?

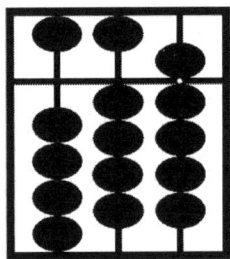 Lesson 13.
What is the 2 digit number?

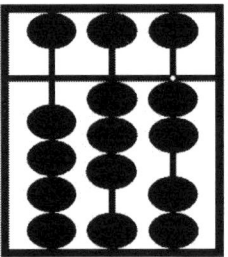 Lesson 13.
What is the 2 digit number?

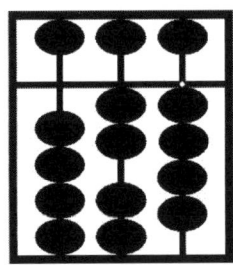 Lesson 13.
What is the 2 digit number?

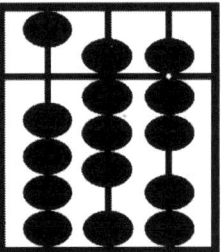 Lesson 13.
What is the 2 digit number?

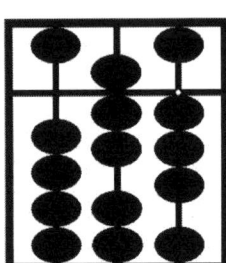 Lesson 13.
What is the 2 digit number?

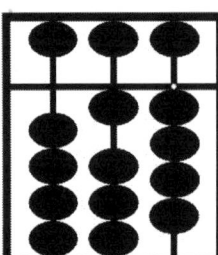 Lesson 13.
What is the 2 digit number?

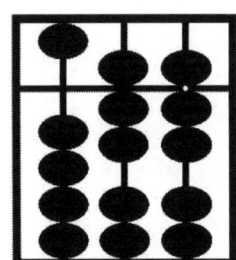
Lesson 13.
What is the 2 digit number?

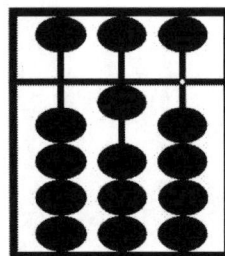
Lesson 13.
What is the 2 digit number?

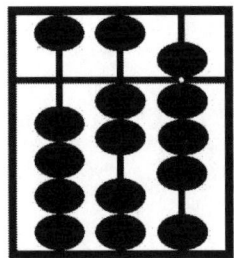
Lesson 13.
What is the 2 digit number?

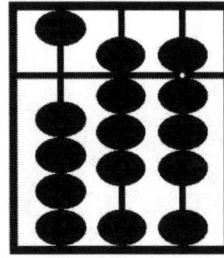
Lesson 13.
What is the 2 digit number?

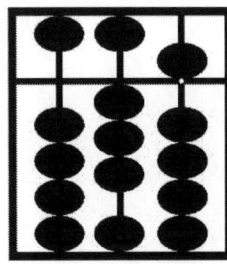
Lesson 13.
What is the 2 digit number?

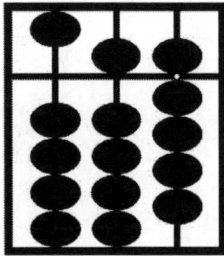
Lesson 13.
What is the 2 digit number?

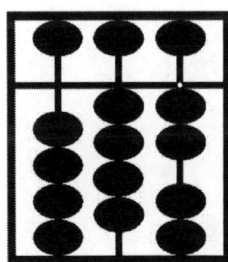
Lesson 13.
What is the 2 digit number?

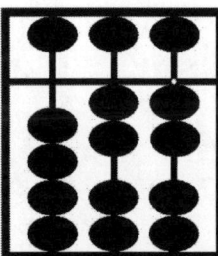
Lesson 13.
What is the 2 digit number?

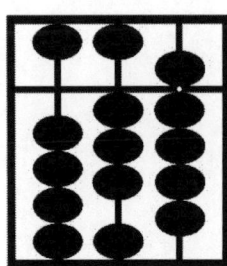
Lesson 13.
What is the 2 digit number?

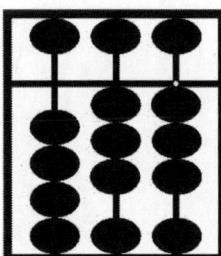
Lesson 13.
What is the 2 digit number?

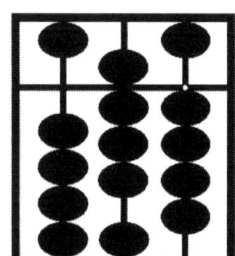
Lesson 13.
What is the 2 digit number?

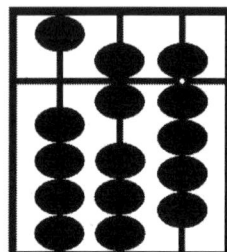
Lesson 13.
What is the 2 digit number?

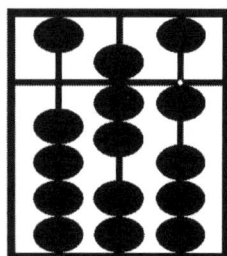
Lesson 13.
What is the 2 digit number?

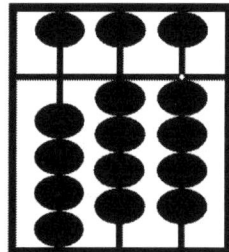
Lesson 13.
What is the 2 digit number?

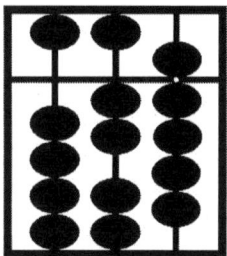
Lesson 13.
What is the 2 digit number?

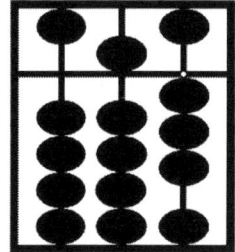
Lesson 13.
What is the 2 digit number?

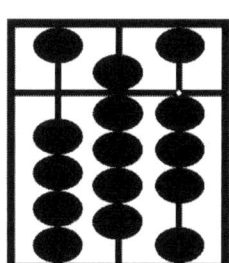
Lesson 13.
What is the 2 digit number?

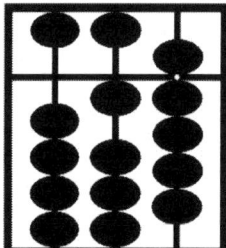
Lesson 13.
What is the 2 digit number?

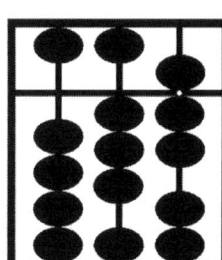
Lesson 13.
What is the 2 digit number?

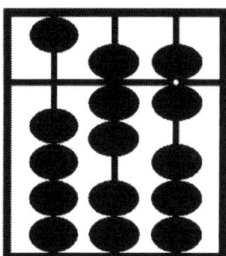
Lesson 13.
What is the 2 digit number?

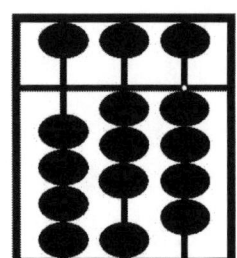
Lesson 13.
What is the 2 digit number?

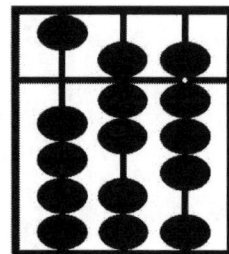
Lesson 13.
What is the 2 digit number?

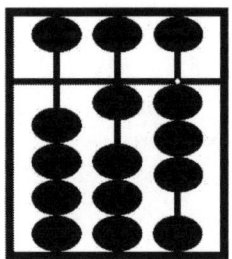
Lesson 13.
What is the 2 digit number?

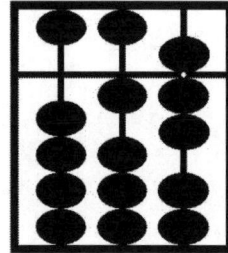
Lesson 13.
What is the 2 digit number?

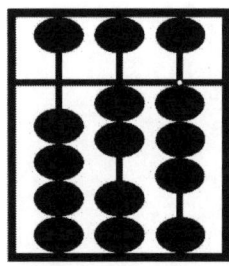
Lesson 13.
What is the 2 digit number?

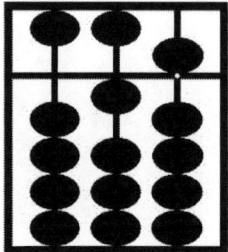
Lesson 13.
What is the 2 digit number?

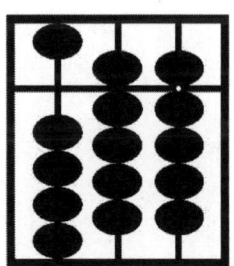
Lesson 13.
What is the 2 digit number?

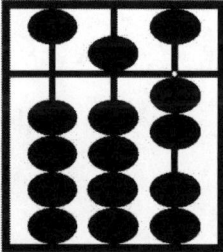
Lesson 13.
What is the 2 digit number?

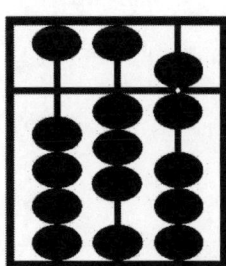
Lesson 13.
What is the 2 digit number?

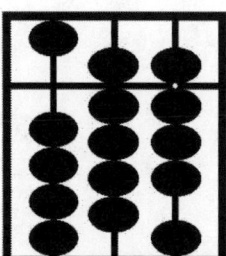
Lesson 13.
What is the 2 digit number?

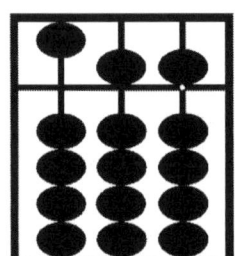
Lesson 13.
What is the 2 digit number?

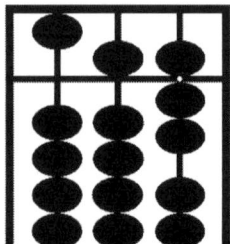
Lesson 13.
What is the 2 digit number?

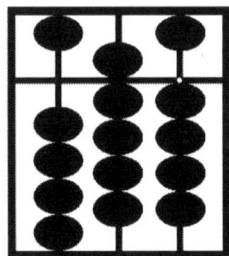
Lesson 13.
What is the 2 digit number?

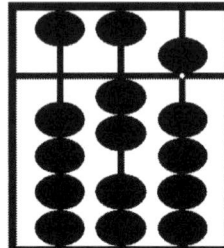
Lesson 13.
What is the 2 digit number?

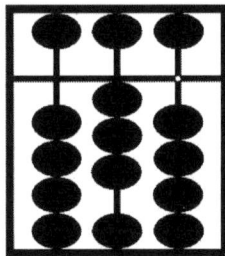
Lesson 13.
What is the 2 digit number?

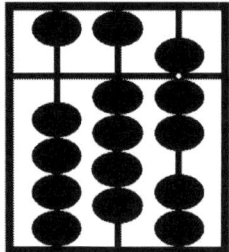
Lesson 13.
What is the 2 digit number?

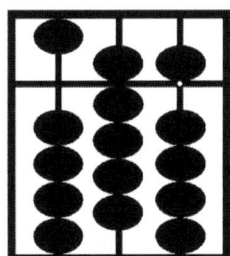
Lesson 13.
What is the 2 digit number?

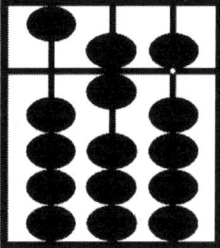
Lesson 13.
What is the 2 digit number?

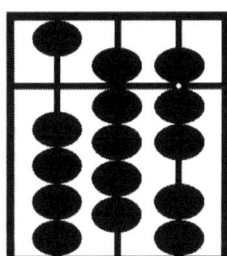
Lesson 13.
What is the 2 digit number?

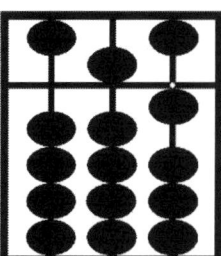
Lesson 13.
What is the 2 digit number?

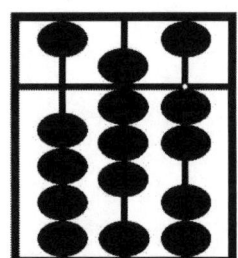
Lesson 13.
What is the 2 digit number?

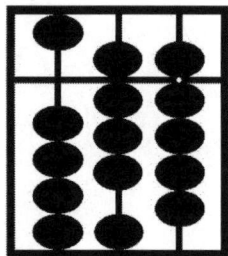
Lesson 13.
What is the 2 digit number?

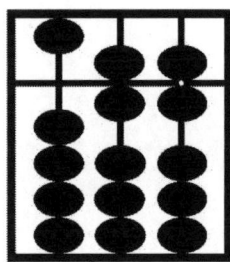
Lesson 13.
What is the 2 digit number?

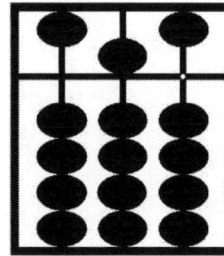
Lesson 13.
What is the 2 digit number?

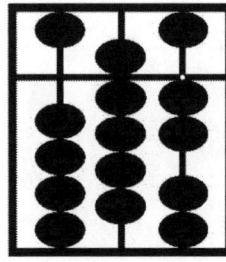
Lesson 13.
What is the 2 digit number?

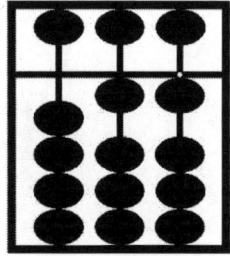
Lesson 13.
What is the 2 digit number?

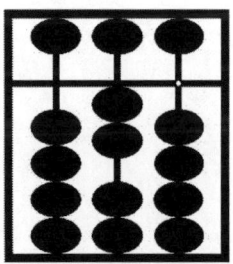
Lesson 13.
What is the 2 digit number?

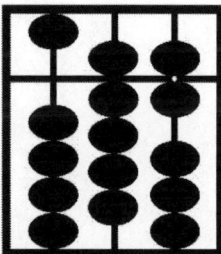
Lesson 13.
What is the 2 digit number?

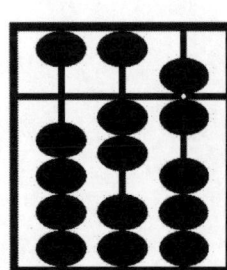
Lesson 13.
What is the 2 digit number?

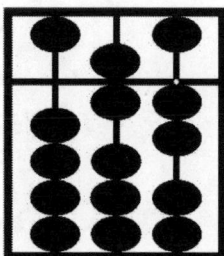
Lesson 13.
What is the 2 digit number?

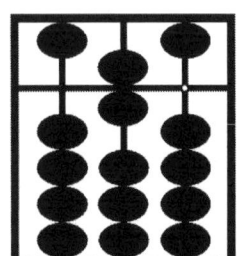
Lesson 13.
What is the 2 digit number?

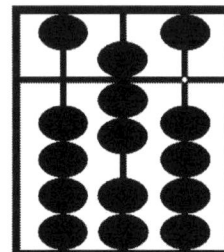
Lesson 13.
What is the 2 digit number?

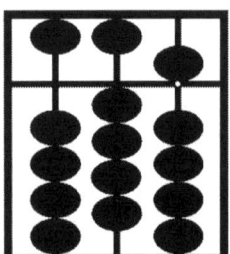
Lesson 13.
What is the 2 digit number?

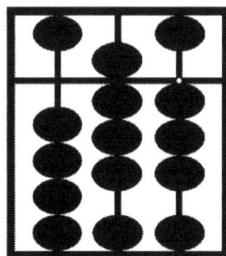
Lesson 13.
What is the 2 digit number?

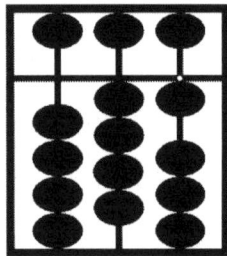
Lesson 13.
What is the 2 digit number?

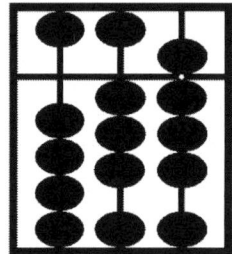
Lesson 13.
What is the 2 digit number?

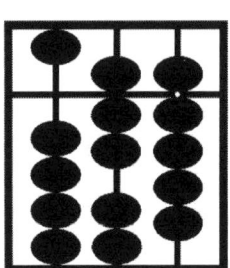
Lesson 13.
What is the 2 digit number?

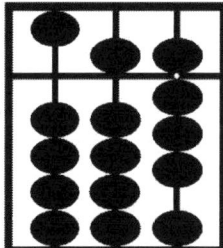
Lesson 13.
What is the 2 digit number?

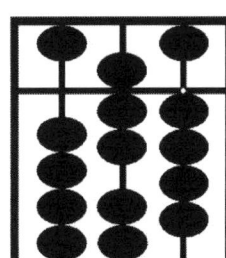
Lesson 13.
What is the 2 digit number?

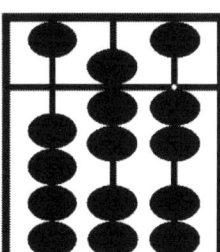
Lesson 13.
What is the 2 digit number?

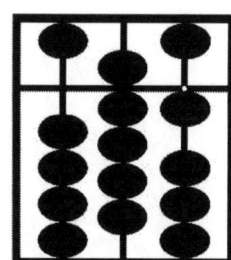
Lesson 13.
What is the 2 digit number?

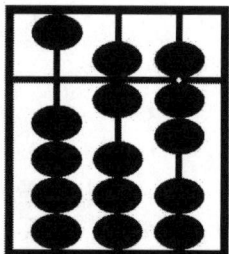
Lesson 13.
What is the 2 digit number?

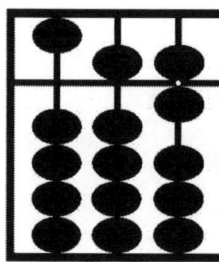
Lesson 13.
What is the 2 digit number?

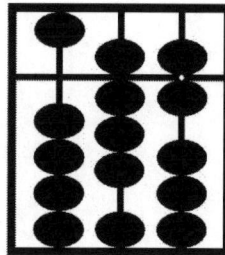
Lesson 13.
What is the 2 digit number?

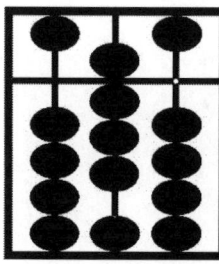
Lesson 13.
What is the 2 digit number?

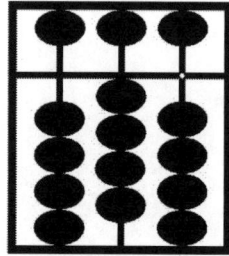
Lesson 13.
What is the 2 digit number?

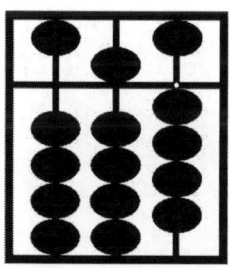
Lesson 13.
What is the 2 digit number?

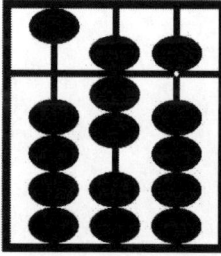
Lesson 13.
What is the 2 digit number?

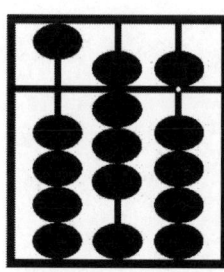
Lesson 13.
What is the 2 digit number?

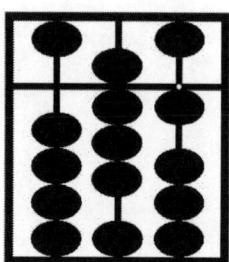
Lesson 13.
What is the 2 digit number?

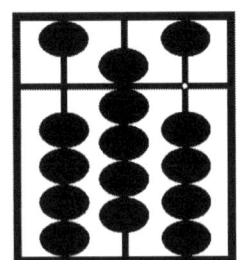
Lesson 13.
What is the 2 digit number?

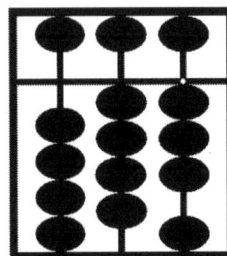
Lesson 13.
What is the 2 digit number?

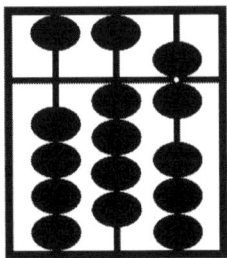
Lesson 13.
What is the 2 digit number?

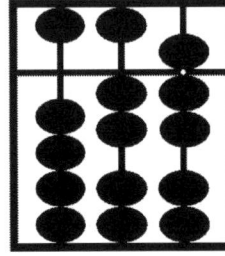
Lesson 13.
What is the 2 digit number?

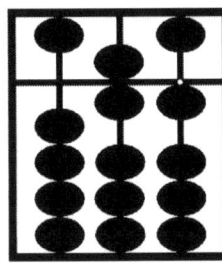
Lesson 13.
What is the 2 digit number?

Answer Key: 68, 12, 18, 16, 63, 64, 21, 48, 31, 49, 32, 24, 87, 73, 14, 77, 10, 28, 88, 35, 59, 42, 22, 39, 33, 84, 69, 71, 44, 29, 53, 93, 19, 37, 76, 34, 78, 13, 17, 23, 15, 99, 52, 36, 98, 55, 57, 94, 25, 30, 47, 95, 65, 97, 51, 82, 89, 66, 50, 92, 11, 20, 96, 26, 62, 60, 70, 45, 83, 41, 38, 79, 58, 74, 72, 91, 67, 56, 86, 80, 40, 54, 75, 85, 81, 90, 43, 46, 27, 61.

Congratulations!

You've successfully counted from 1 to 99 on the abacus!

Chapter Review

1. Always start on **ones place value**, where the period is.
2. **Tens place value** is to the left of **ones place value.**
3. Add 1, 2, 3 & 4 with your thumb and subtract with your index finger.
4. Add 5 with your index finger and subtract with your index finger.
5. Add 6, 7, 8 & 9 with your thumb and index at the same time and subtract with both fingers as well.
6. Perform the clear action anytime the abacus needs a reset.
7. Always hold the abacus with the off-hand so it will not shake when you're working on it.
8. Use your writing hand to add and subtract numbers on the abacus.
9. The other fingers that are not used should be tucked in like a fist.

Chapter 6: Rules of Addition & Subtraction

Next, we will need to learn the rules of addition and subtraction on the abacus. We will work your way up slowly as there are a total of 34 rules that you will need to memorize by heart. However, there is no rush in memorizing all the rules at once as we will go over them one-by-one as we progressively advance in the abacus.

The rules are what makes the abacus fun. For instance, when you have a 4 on your abacus and you want to add another 4, there is no room for another 4 beads. The rules are another way to substitute the numbers into groups. For this example, +4=+5-1, meaning to add 4 on the abacus, you will have to add 5 and take away 1, and you will have your answer: 8. In this book, we will not be practicing any questions with rules, yet. We will focus on your basic foundation of addition and subtract first. However, here is the rule sheet for your reference.

For your exercise, work on 1 page per day. Follow the directions on top of the page to learn at your own pace. Each page of exercise will have new materials to learn.

R U L E S

Addition

Composition of 10		Composition of 5	
Number(s) the rule applies to	Rule	Number(s) the rule applies to	Rule
9	$+1=-9+10$ [1]	4	$+1=+5-4$ [10]
8,9	$+2=-8+10$ [2]	3,4	$+2=+5-3$ [11]
7,8,9	$+3=-7+10$ [3]	2,3,4	$+3=+5-2$ [12]
6,7,8,9	$+4=-6+10$ [4]	1,2,3,4	$+4=+5-1$ [13]
5,6,7,8,9	$+5=-5+10$ [5]		
4,9	$+6=-4+10$ [6]	5,6,7,8	$+6=+1-5+10$ [14]
3,4,8,9	$+7=-3+10$ [7]	5,6,7	$+7=+2-5+10$ [15]
2,3,4,7,8,9	$+8=-2+10$ [8]	5,6	$+8=+3-5+10$ [16]
1,2,3,4,6,7,8,9	$+9=-1+10$ [9]	5	$+9=+4-5+10$ [17]

Subtraction

Decomposition of 10		Decomposition of 5	
Number(s) the rule applies to.	Rule	Number(s) the rule applies to.	Rule
10	$-1=-10+9$ [18]	5	$-1=+4-5$ [27]
10,11	$-2=-10+8$ [19]	5,6	$-2=+3-5$ [28]
10,11,12	$-3=-10+7$ [20]	5,6,7	$-3=+2-5$ [29]
10,11,12,13	$-4=-10+6$ [21]	5,6,7,8	$-4=+1-5$ [30]
10,11,12,13,14	$-5=-10+5$ [22]		
10,15	$-6=-10+4$ [23]	11,12,13,14	$-6=-10+5-1$ [31]
10,11,15,16	$-7=-10+3$ [24]	12,13,14	$-7=-10+5-2$ [32]
10,11,12,15,16,17	$-8=-10+2$ [25]	13,14	$-8=-10+5-3$ [33]
10,11,12,13,15,16,17,18	$-9=-10+1$ [26]	14	$-9=-10+5-4$ [34]

Download your PDF copy HERE (for ebook).

Abacus 101: Beginner Abacus Math

Chapter 7: Thumb up & Index down

Addition: Thumb up for bottom beads. e.g. 1+1, 3+1, 1+2, 1+3
Subtraction: Index down for bottom beads. e.g. 1-1, 2-1, 3-2, 4-4

No.	1	2	3	4	5	6	7	8	9	10
1	2 2 -1	3 -2 3	1 2 1	4 -3 2	1 2 -1	1 -1 1	2 -2 1	1 3 -2	4 -1 1	4 -2 1
Ans										
2	3 1 -4	2 1 -2	4 -2 2	2 -1 1	1 1 2	3 -3 4	2 2 -2	4 -3 3	1 3 -1	4 -4 1
Ans										
3	4 -3 1	1 3 -4	2 -1 3	1 2 -2	2 -1 2	3 -1 2	4 -4 3	3 1 -2	3 -2 3	2 -2 3
Ans										
4	2 2 -3	1 -1 3	4 -2 1	1 3 -3	3 -3 3	4 -4 4	2 1 -1	1 -1 2	3 -3 1	4 -4 2
Ans										

Abacus 101: Beginner Abacus Math

Chapter 8: Top bead & Index down

Addition: Index down for +5 e.g. 1+5, 2+5, 3+5, 4+5

No.	1	2	3	4	5	6	7	8	9	10
1	1 3 5	2 -1 3	3 -2 5	3 5 -3	4 5 -1	5 4 -2	6 3 -4	5 1 3	6 2 -3	6 -1 4
Ans										
2	1 3 -1	2 5 -2	3 -3 5	4 5 -2	5 4 -3	6 -1 3	7 -2 4	8 -3 1	8 -3 4	8 -2 3
Ans										
3	2 1 -2	3 5 -1	8 -3 2	1 2 1	9 -4 3	2 2 -3	4 5 -4	7 -1 2	2 1 5	1 3 -3
Ans										
4	7 -1 3	6 2 1	1 3 -2	5 3 -1	4 -4 5	3 1 5	7 1 -2	2 -2 5	5 4 -1	1 5 3
Ans										

Copyright © 2017, by David Dong, All Rights Reserved.

Abacus 101: Beginner Abacus Math

Chapter 9: Top bead & Index up

Subtraction: Index up for -5 e.g. 5-5, 6-5, 7-5, 8-5

No.	1	2	3	4	5	6	7	8	9	10
1	6 -5 3	9 -5 5	8 -5 1	7 -5 2	6 2 -5	3 5 -5	2 5 -5	5 -5 4	2 5 -1	7 -5 5
Ans										
2	9 -5 -3	7 2 -5	4 -3 5	7 -5 1	8 -1 -5	8 -5 -2	9 -4 3	6 -5 1	1 5 -5	5 4 -5
Ans										
3	2 5 -2	5 4 -3	3 5 -1	7 2 -5	8 -2 -5	1 2 5	5 3 -5	7 1 -5	4 5 -3	6 1 -5
Ans										
4	7 -1 -5	6 -5 5	4 -4 5	9 -5 -2	8 -5 -1	6 3 -4	3 -1 5	5 4 -4	1 5 3	7 -5 -1
Ans										

Abacus 101: Beginner Abacus Math

Chapter 10: Thumb + Index additions

Addition: Thumb up & Index down for +6 e.g. 1+6, 2+6, 3+6

No.	1	2	3	4	5	6	7	8	9	10
1	2 -1 6	2 6 -2	8 -5 6	3 6 -1	6 2 -5	3 6 -3	8 -3 1	9 -2 -5	7 -5 1	1 2 6
Ans										
2	3 6 -4	2 6 -5	1 6 2	6 -5 3	5 4 -1	1 6 -5	2 6 -3	7 -5 2	3 -1 6	9 -5 -3
Ans										
3	7 -5 6	9 -4 3	2 -2 6	6 -5 1	8 -3 4	4 -1 6	5 3 -2	8 -5 -1	3 -2 5	1 5 2
Ans										
4	2 6 -1	5 1 3	3 6 -2	5 2 1	6 -5 6	6 1 -5	4 -2 6	9 -1 -5	8 -5 1	3 5 -2
Ans										

Copyright © 2017, by David Dong, All Rights Reserved.

Abacus 101: Beginner Abacus Math

Chapter 11: Thumb + Index additions

Addition: Thumb up & Index down for +7 e.g. 1+7, 2+7

No.	1	2	3	4	5	6	7	8	9	10
1	6 -5 7	3 -2 7	2 7 -3	6 1 2	4 -3 7	2 7 -4	4 -2 7	8 -5 6	3 -1 7	2 -2 5
Ans										
2	5 -5 7	7 -5 6	1 7 -2	4 -1 6	9 -3 1	5 -5 4	2 -2 7	7 -1 -5	2 6 -5	8 -5 -2
Ans										
3	5 2 1	9 -5 -3	6 -5 6	5 4 -2	3 6 -5	1 7 -5	8 -5 -1	4 -4 7	2 -1 7	8 -3 4
Ans										
4	4 -4 6	9 -5 -1	5 -5 2	6 -5 1	4 -2 6	2 6 -3	3 -3 7	1 6 -2	8 -5 1	1 -1 7
Ans										

Copyright © 2017, by David Dong, All Rights Reserved.

Abacus 101: Beginner Abacus Math

Chapter 12: Thumb + Index additions

Addition: Thumb up & Index down for +8 e.g. 1+8

No.	1	2	3	4	5	6	7	8	9	10
1	1 8 -3	1 8 -1	3 -2 7	4 -4 9	2 7 -5	3 -2 8	6 -5 6	8 1 -2	7 -5 1	9 -4 4
Ans										
2	6 -5 8	3 -3 9	4 5 -1	8 1 -3	3 6 -5	6 -5 7	2 -1 8	5 -5 9	4 -3 2	7 -2 4
Ans										
3	4 -3 8	2 -1 7	3 6 -4	9 -4 3	1 7 -2	9 -3 -5	6 -5 1	2 7 -4	3 -2 6	5 4 -1
Ans										
4	7 2 -5	8 -5 1	4 -3 6	1 8 -4	3 -1 2	7 -5 6	2 -1 5	9 -2 -1	6 3 -4	1 -1 9
Ans										

Copyright © 2017, by David Dong, All Rights Reserved.

Abacus 101: Beginner Abacus Math

Chapter 13: Thumb + Index subtractions

Subtraction: Thumb down & Index up for -6 e.g. 6-6, 7-6, 8-6, 9-6

No.	1	2	3	4	5	6	7	8	9	10
1	6 -6 4	7 -6 3	8 -6 2	9 -6 1	3 5 -6	5 4 -6	2 -1 8	1 8 -6	7 -2 4	2 5 -6
Ans										
2	3 5 -2	2 7 -6	7 -6 8	4 -4 9	8 1 -6	5 2 1	9 -3 -5	6 -5 8	1 -1 9	7 2 -6
Ans										
3	9 -6 -2	8 -6 5	5 2 -6	6 3 -1	9 -1 -6	4 -4 7	7 -6 1	6 -6 9	2 -1 3	8 -5 -2
Ans										
4	7 2 -5	3 -1 7	8 -6 1	9 -2 -6	5 4 -5	1 8 -3	2 7 -4	4 5 -2	6 1 -6	4 -3 8
Ans										

Abacus 101: Beginner Abacus Math

Chapter 14: Thumb + Index subtractions

Subtraction: Thumb down & Index up for -7 e.g. 7-7, 8-7, 9-7

No.	1	2	3	4	5	6	7	8	9	10
1	8 -7 6	2 6 -5	1 8 -7	6 3 -7	9 -7 2	4 -1 6	5 4 -3	9 -6 -1	8 -7 5	7 -7 9
Ans										
2	3 5 -2	8 -7 3	7 -6 1	9 -7 6	2 6 -7	4 -3 8	5 2 -7	6 1 -2	1 8 -6	2 7 -5
Ans										
3	8 -7 1	9 -6 5	5 2 -6	3 5 -7	1 8 -4	6 3 -7	4 -2 6	8 -7 2	7 1 -3	5 4 -7
Ans										
4	8 -7 8	9 -7 5	1 8 -3	2 5 -6	7 -7 4	3 5 -6	4 5 -7	6 3 -2	5 3 -7	8 -6 1
Ans										

Copyright © 2017, by David Dong, All Rights Reserved.

Abacus 101: Beginner Abacus Math

Chapter 15: Thumb + Index subtractions

Subtraction: Thumb down & Index up for -8 e.g. 8-8, 9-8

No.	1	2	3	4	5	6	7	8	9	10
1	5 4 -8	1 8 -9	7 2 -8	9 -9 3	3 6 -4	8 1 -2	2 7 -6	4 5 -8	6 3 -5	9 -9 7
Ans										
2	8 1 -7	9 -8 6	4 5 -3	8 -6 2	5 -5 9	3 -2 8	7 -7 4	1 8 -5	9 -9 1	1 8 -3
Ans										
3	6 3 -9	1 8 -7	7 -6 3	3 5 -2	5 4 -6	8 -7 5	9 -9 8	5 2 -1	7 1 -6	2 -2 4
Ans										
4	2 5 -6	8 -2 3	1 6 -2	7 -5 7	8 -8 4	4 -3 1	9 -8 5	6 -5 8	3 6 -9	8 -8 5
Ans										

Beginner Abacus Math by Tong Dazai

Made in the USA
Monee, IL
03 May 2026